Dedicated to the people who strive to solve world's energy problems

https://www.ucsusa.org/energy

Text by

Gavin Krebs

Bob Jia

Dr. Ellen Cavanaugh

Mr. Jake Gorczyca

Cartoon images created in Vyond.

ISBN: 9 781716 016561

Grow a Generation
Sewickley, PA 15143
www.growageneration.com

Any and all profits from the sale of this book benefit charity

Mr. W was teaching his science class saying, "Few inventions have more impact on your daily life as the battery."

"Really?" two of his best students, John and Akira, said aloud.

Correct Akira. The history of batteries is fascinating. Batteries store chemical energy that is converted into electricity. In 1800, Alessandro Volta demonstrated the first battery. He stacked discs of copper (Cu) and zinc (Zn) separated by cloth soaked in salty water.

"I want to make a battery." - Akira said

"Me too!" - Added John

This battery came to be known as the "Voltaic Pile" and generated an electric current when electrons were released from the zinc into the copper.

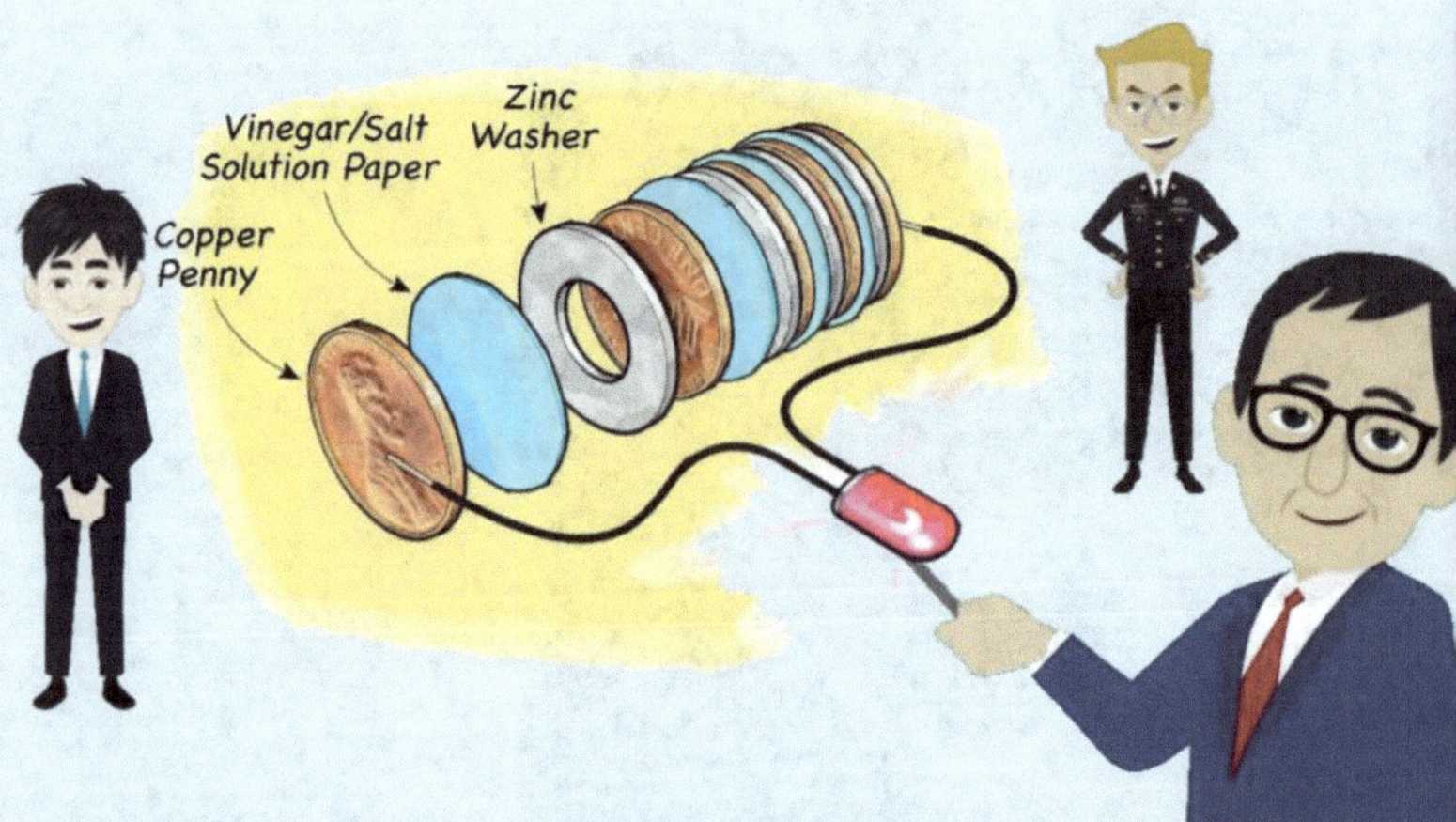

Vinegar/Salt Solution Paper

Zinc Washer

Copper Penny

Ions inside the battery release electrons through the anode. The electrons travel outside the battery in a circuit, drawn to the cathode, usually stopping on the way to light an LED or spin a motor.

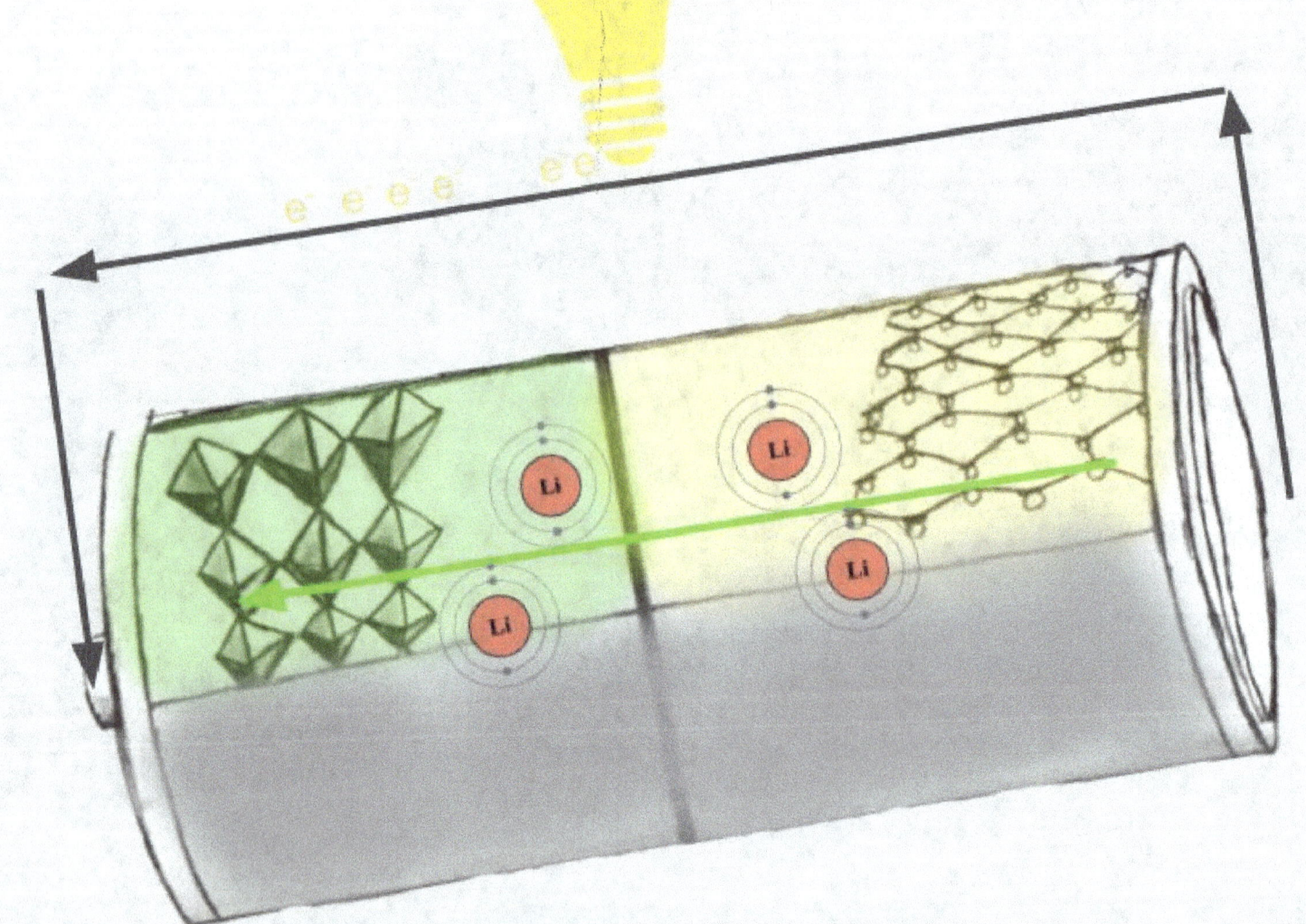

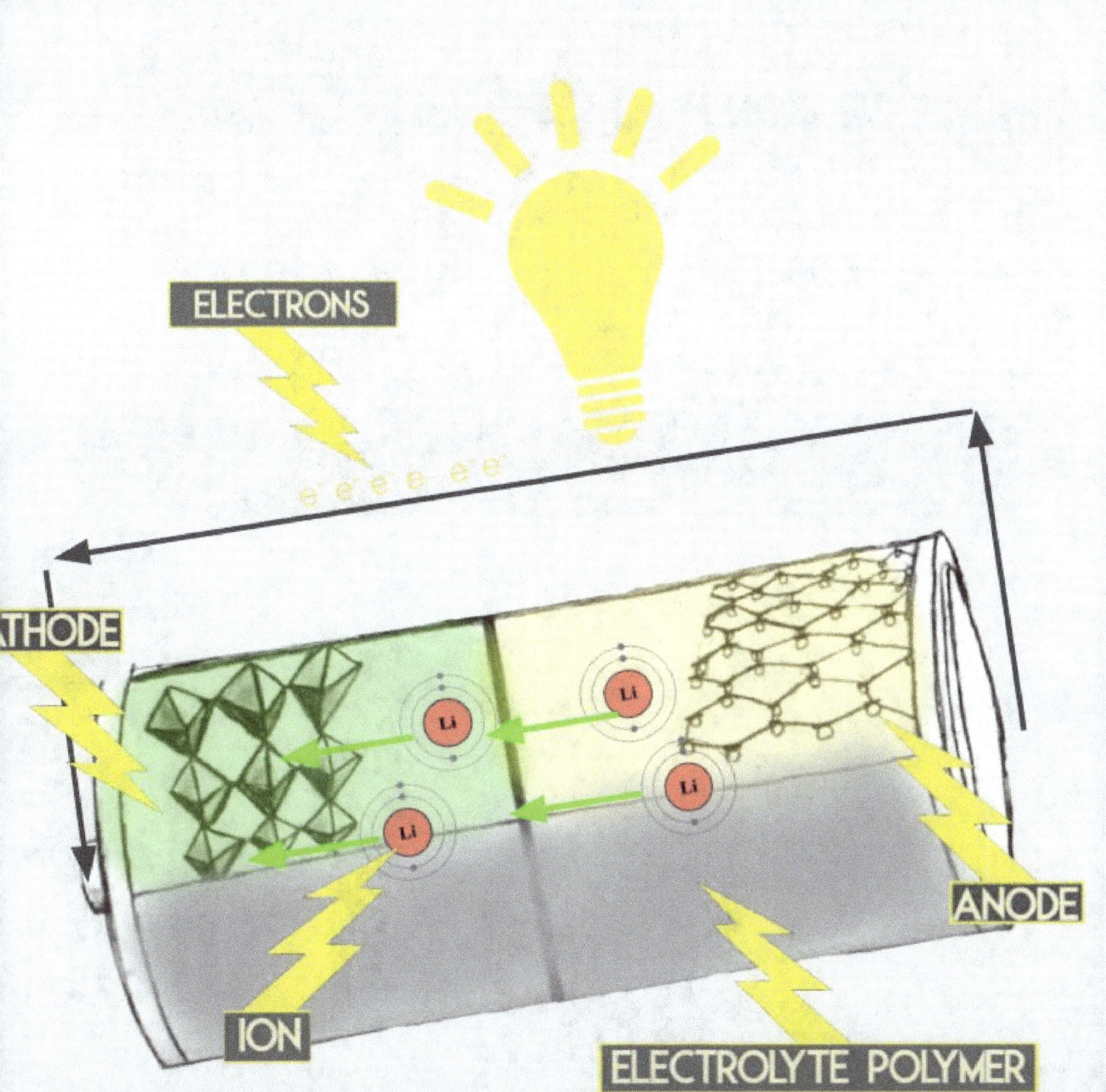

Imagine a sled on top of a hill, it is just like the ions inside a fully charged battery. The sled wants to get to the bottom of the hill just like how the _ions_ in the _anode_ of the battery want to get to the _cathode._ The sled travels down the hill and across the snow until it gets to the bottom just like how the ions in a battery travel across the _electrolyte_ through the _barrier_ to the cathode.

Once the sled is at the bottom of the hill, it needs to be pushed back up to the top in order to be used again. Just like the sled, the ions in the battery need to be pushed back over to the anode from the cathode in order to use the battery again. This is done when the battery is charged.

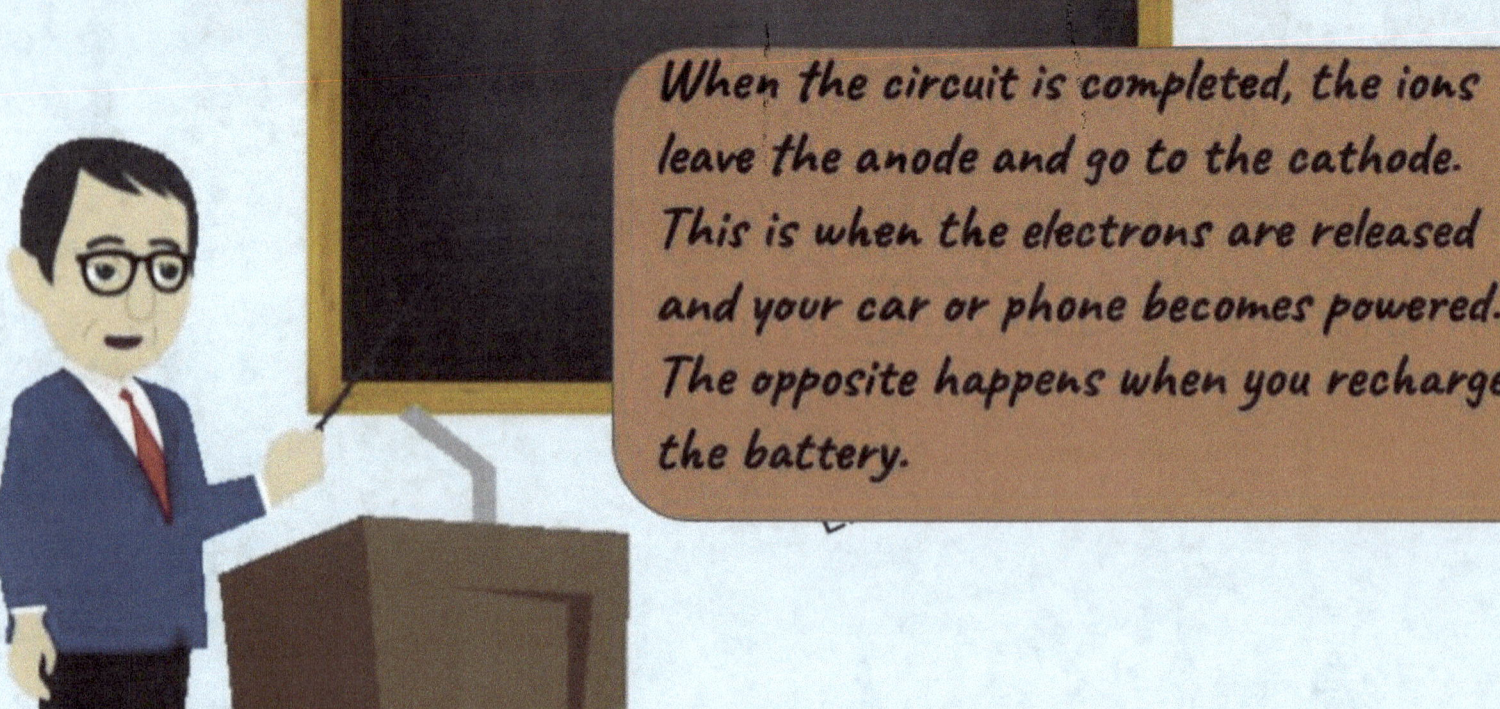

Whittingham decided to add lithium to his battery, but did not stabilize the lithium with any other element. This led to the battery catching on fire. Whittingham then added aluminum to stabilize the lithium.

Goodenough knew a lot about battery cathodes so, he set out to make a better one. Goodenough added a cobalt-oxide cathode that could hold more lithium ions than previous cathodes which also had a tendency to cause the battery to fail and catch on fire. The reason why the cathode had a greater capacity was because it could hold more ions by intercalating them. This ultimately meant that the ions were able to be grouped more tightly together in a small area due to the rearranged structure. The old metal sulfide cathode did not intercalate the ions, so it held less ions. When there are more ions, there are more electrons (negatively charged particles used to power electronic devices).

Yoshino knew a lot about anodes so, he set out to make a better battery anode. Yoshino reinvented the lithium metal anode and replaced it with a much more stable carbon-rich material. Because the material was more stable, it was also safer. This anode was able to hold more lithium ions because, like the cathode, it intercalated the ions. By containing more lithium ions, the anode could now hold more electrons, and therefore, hold more electrical potential energy.

Yoshino and Whitingham Fail to win the science fair. The two boys were very upset that neither one was able to win. But Whittingham had an idea.

And so the two young scientists set about creating the battery that would revolutionize technology for decades to come.

Together they entered the Science fair the following year.

Goodenough and Yoshino shared their new battery. It was amazing and it was the best rechargeable battery ever built. They won the science fair because they all worked together.

When put together, the two parts of the battery that Yoshino and Goodenough made were able to function well together and make the battery the best battery ever made. The same battery they made almost 30 years ago is still the best battery today. Only minor changes to the battery have been made to make the best battery today and that alone demonstrates the significance of this invention.

In 1991, a major Japanese electronics company started selling the first lithium-ion batteries, leading to a revolution in electronics. Mobile phones shrank, computers became portable and MP3 players and tablets were developed. Therefore, the demand for high capacity and high voltage batteries surged.

Coincidently, the lithium-ion battery has been changed and improved; among other things, John Goodenough has replaced the cobalt oxide with iron phosphate, which makes the battery more environmentally friendly.

It is very important to understand that when people work together, they can develop technology more quickly. Even though working together does make allow for rapid innovation, these advancements do not occur all at once. There are times where new ideas will fail over and over, but as long as the inventors remain determined to make their innovation a reality, a great creation will result.

Without this invention, we would not have electric cars, cell phones, laptops, or other portable electronic devices. Because of this great achievement, we were able to enjoy 15 years of improved technology. Now, we just need to make the next battery breakthrough.

Cutting Edge Battery Research

Relative to a conventional lithium-ion battery, solid-state lithium-metal battery technology has the potential to increase the cell energy density (by eliminating the carbon or carbon-silicon anode), reduce charge time (by eliminating the charge bottleneck resulting from the need to have lithium diffuse into the carbon particles in conventional lithium-ion cell), prolong life (by eliminating capacity fade that results from the unwanted chemical side reaction between the carbon and liquid electrolyte in conventional lithium-ion cells), improve safety (by eliminating the combustible organic porous separator and organic anolyte material in conventional cells) and lower cost (by eliminating the anode materials and manufacturing costs).

Quantumscape

QuantumScape is a company that has been around since 2010 and has been committed to researching battery technology since. The company is currently aiming to develop a battery with an energy density of 1,000 Wh/L, with the intent of using the battery in electric vehicles.

Such an energy-density would translate to approximately 50-80% more range/charge vs today's production electric vehicles. The company's future is bright with backing from major investors including Bill Gates and major automotive corporations including Volkswagen.

TDK Electronics

Lithium-ion batteries have nearly peaked, which means we must pursue new powercell technology to power electric vehicles. TDK is shooting to replace lithium-ion batteries with solid-state batteries in the next few years. With hopes to release the first commercially viable solid state battery in the next year, TDK is building upon the revolutionary technology they have already developed.

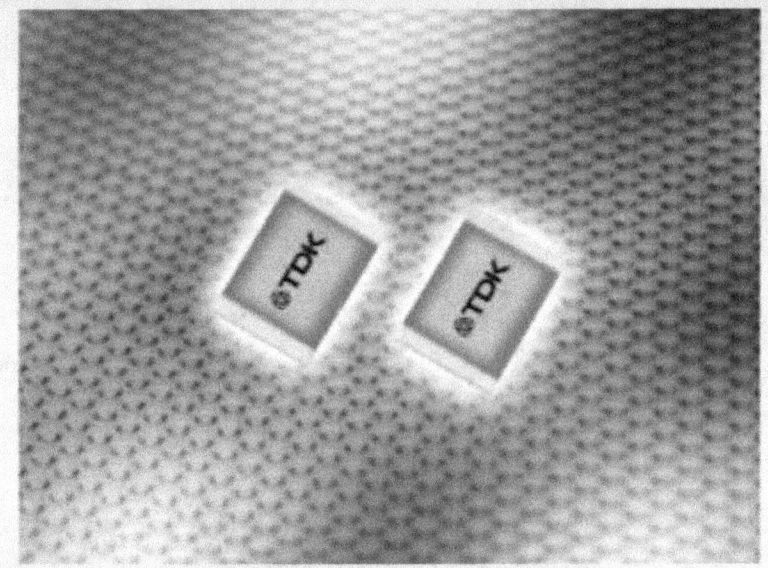

Pictured above: TDK's rechargeable solid
state SMD battery.

TDK has already created the first solid-state SMD battery, using their ceramic electrolyte. Their new solid state powercell would utilize a ceramic-based electrolyte developed by TDK. Technology of this sort teases the possibility of travelling over 1,000 miles per charge for EV's that utilize the technology.

Selfless, dedicating countless hours connecting students with professional mentors, Dr. Ellen Cavanaugh was the supportive, and innovative founder of the Baden STEM Leadership Center. She taught students how to properly conduct research and interact with professionals so they could learn new information and find their passion. Always energetic, encouraging thousands of students to go the extra mile and never hold back from learning as much as they possibly can. Dr. Ellen's dream has always been to help others achieve greater levels of excellence, and we are grateful for the lasting, positive impact on the lives of her students and their families.